Building or Renovating a Small Barn for Your Horse

Jackie Clay

CONTENTS

Introduction

When it comes to providing adequate shelter for domestic animals, horses are among the easiest you'll deal with. Contrary to what you might think, horses are sturdy, adaptable creatures, and their housing requirements are very basic: shade, ventilation, protection from the elements, and comfortable, dry footing. Now, this may sound simple, and it can be, but remember that horses are also large, strong, active, and playful. I've seen a horse happily and safely housed in a three-sided shed that cost less than $200 to build — and I've seen horses sick with respiratory diseases, thrush, navicular disease, lameness, and severe injuries, all of which occurred in shelters that were very expensive and looked picture perfect.

Building a horse barn is quite a project — and there is a vast and varied array of plans and books available to help you. But renovating a shelter to make it fit for just one or two horses, or building a sturdy three-sided shelter, can be quite simple, even if you have only basic carpentry skills. And that's what I'll talk about here.

The Fundamentals of Building for Horses

The fundamentals of proper construction for horse housing are:

- **Stay on high ground.** Make sure that the shelter is located on a slight rise or hill, which will ensure adequate drainage away from the structure. Horses need dry footing; otherwise, such foot diseases as thrush and grease heel will result. You wouldn't want to stand in constant mud during a rainy spell, right? Well, neither does your horse.
- **Overbuild rather than underbuild.** A lot of horses are injured kicking through light plywood or sheet-metal walls while playing, knocking 2 x 4s loose, tearing their hides on exposed nails, and — believe it or not — pushing apart walls while scratching their tails. Unlike the case of a garden shed or even a garage, 2 x 4s aren't adequate for horse housing. You'll need to use 4 x 4s or 2 x 6s or greater. I've found that using rough-cut sawmill lumber adds a greater dimension of strength and safety — and costs a lot less.

- **Build a sturdy understructure.** The timber-frame and pole-barn types of construction are the sturdiest, and thus the best choices for horse barns. If the structure you're renovating uses conventional framing with 2 x 4 posts, you'll need to add extra structural support to make the building strong enough to survive years of use by playful, rambunctious horses.
- **Use heavy-duty hardware — hinges, latches, and so on.** Even if you're working with a tight budget, this is no place to scrimp. It's better to employ a set of used, auction-bought hinges than a cheap new store-bought set! Always think heavy, strong, and safe when you're planning, buying material, and building. It's much cheaper to build a safe, long-lasting building or stall than it is to construct a low-priced structure that requires constant repair and remodeling due to horse damage. And it won't take 10 minutes longer to build!

Tips and Techniques for Remodeling

Often the best and most obvious choice for horse housing is simply to modify an existing structure. Unused older cow barns, chicken coops, wooden sheds, and garages can often easily and economically be remodeled to suit a horse. And you don't have to be a professional contractor to build good, solid housing for your horse. Remember, horses can live well in structures as simple as lean-tos and three-sided run-in sheds.

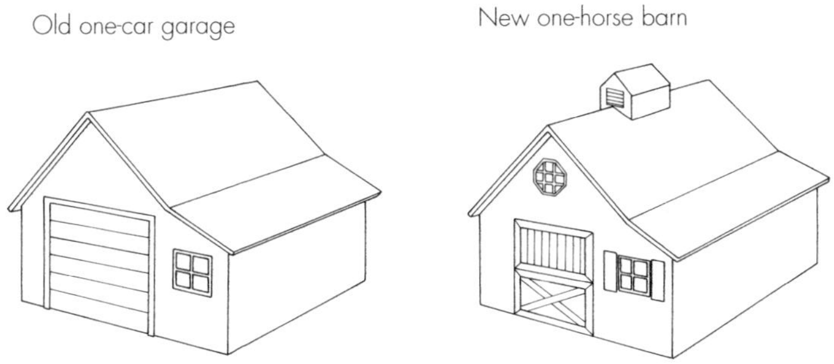

Almost any small structure that's still in good condition can be transformed into a shelter for your horse.

What's a Run-in Shed?

A run-in shed is a three-sided shelter that allows horses to come and go as they please, taking advantage of the shed's shade in the heat of the day and the sun's warmth during cold winters. The three sides offer shelter from the wind, no matter which direction it's blowing from; the sloping roof protects against rain, sleet, and snow.

A three-sided run-in shelter

Considerations for Reuse

If you're considering modifying an existing structure, here are a few things to keep in mind:

- **Is the structure in the right place?** Is the building set on an elevation so that water will drain away from it? Is it near or in the pasture you intend to use? Does it open to the south? I like a barn to run east–west, facing north, with paddock or pasture access on its southern side. This gives the horses additional protection from sharp, cruel north winds as well as a sunny spot to relax in the afternoon.
- **Is there enough headroom for the horse?** A building with less than 7 feet (2.1 m) of headroom is not a good choice for horse housing. (When calculating headroom, measure the distance from floor to the rafters or support beams that tie the walls together.) Most horses will at some time or another get playful and throw up their heads. You wouldn't want an animal to get injured by hitting rafters or beams.

- **Is the structure strong enough?** Usually an old cow barn is well built, with strong, durable timbers and heavy construction. But the smaller buildings you might be considering, such as chicken coops and small sheds, are often built with 2 x 4s, which aren't strong enough to house a horse without extra beefing up.
- **Is the structure big enough?** You should provide an ample stall — about 10' x 12' (3 m x 3.7 m) — for your horse. Anything larger allows for more exercise inside but takes longer to clean and bed — besides being more costly to keep in bedding. If two horses share the same paddock or pasture, remember that horses "buddy." Often both will occupy a single stall, even when you provide two separate ones, so the stall should be large enough to accommodate a pair (about 10' x 14', or 3 x 4.3 m). Housing for ponies can, of course, be proportionately smaller.
- **Is the structure safe?** Don't, for example, use a corner of a tractor shed or garage for horse housing. Many more fires occur in structures used for equipment and vehicles than occur in livestock shelters. Gasoline and other combustible materials are never a safe mix with horses. I've seen garages burn from a ruptured fuel line dripping gas onto a hot manifold, a backfiring carburetor, and a welder sending sparks into flammable material. For the safety of your animals and your own peace of mind, you don't want to house your horses where you'll have that sort of activity occurring.
- **Is the framework for the roof in good condition?** If your structure's roofing needs replacing, it's simple enough to secure new sheets of corrugated metal. However, if the boards forming the frame that holds up the roof of your structure aren't strong and secure, you'll need to replace them — a potentially expensive proposition.
- **Does the existing structure have a cement floor?** Cement is very hard on a horse's feet and joints, not to mention uncomfortable to lie on. If you choose to use a cement-floored structure, you need to bed your animal with at least 6 inches (15 cm) of sand or sawdust, topped with removable bedding like straw or shavings.

Four Rules for Renovating Horse Shelters

Remodeling considerations depend on the type of care you want to give your horse. The easiest and often the happiest solution is to transform the structure into a stall that's available to your animal at all times. Such a run-in shed allows the horse to be either at pasture, in a paddock, or safely in his stall when a sudden storm comes up or flies become bad. He isn't dependent on you to bring him inside or let him out for exercise or pasturing.

Whether you're renovating an existing structure into a run-in or an enclosed shed, however, follow these basic rules:

1. Reinforce the frame. Unless you're working with sturdy pole-building or timber-frame construction that's still in good shape, you'll need to reinforce the corner posts that frame the structure. Use a posthole digger to dig a hole 3½ feet (1 m) deep on the inside corners. If the floor is concrete, you'll first have to knock out a chunk where each inside corner will be, using a sledgehammer (and, of course, wearing eye protection). Into this hole, secure a 6 x 6 post that runs from the ground to a cross beam or rafter. Bolt the post to the existing corner post and use a fresh batch of cement to set it in place. This post will be very strong, firmly anchoring the inside wall of the structure. In a similar manner, set an anchor pole in each corner of the structure and on each side of the door, if there is one. For additional strength, set more reinforcement poles as necessary so that no pole is more than 12 feet (3.7 m) from another.

Most outbuildings are constructed with 2 x 4 bracing set on a sill plate, which is set on concrete footing. To make this framework strong enough to shelter a horse, nail reinforcing studs to each 2 x 4 and to the sill plate. You can even add bracing between the 2 x 4s in high-traffic areas.

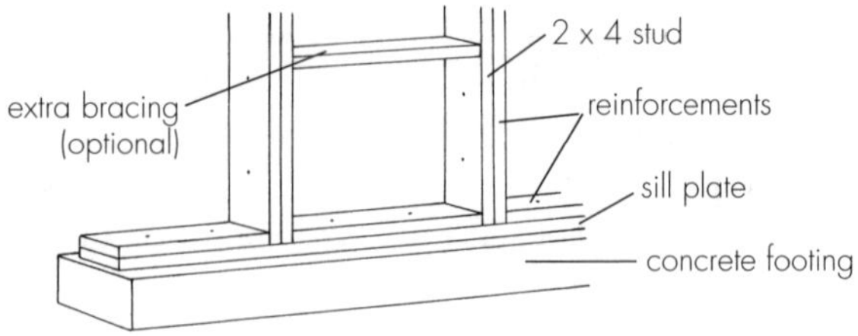

In addition, if the existing structure was built with 2 x 4 studs and sill plates, you'll need to double up, or add reinforcing members to each stud and section of sill plate. I've even done this using rough-sawed (hewn with a chain saw) logs from our woods, flattened on two or more sides. Nail or screw (using a power screwdriver) each reinforcement member into its matching 2 x 4 stud.

2. Reinforce the walls. If a run-in shed or any of the areas in an enclosed shed with which a horse might come in contact doesn't already have sturdy, 2-inch-thick (5 cm) walls with sturdy planking on the inside of the stall, you'll need to redo the planking (see below). If you want to use plywood or steel sheeting as your siding material, do so on the outside where it won't be exposed to the horse, his activity, and his playfulness.

3. Examine the roof. If the roof is not secure and watertight, you'll need to repair or replace it — see page 23. In addition, examine the framework that supports the roof. If you're working on an older building, this support structure may have fallen into disrepair and be in need of reinforcement or replacement.

4. Provide for adequate light and ventilation. If you're working on an enclosed shed, be sure there's at least a small window in the stall for light and ventilation, as well as extra sky panels and windows to allow light to enter the building. This is especially important if you don't have electricity to power artificial lighting in the barn. In the stall, it's best to have a high window about 12 to 18 inches (30–45 cm) tall and 2 feet (24 inches) wide that you can slide open as desired.

Planking a Run-In Shed

Once you've framed a run-in shed with reinforcing posts, you can start the planking. The minimum thickness of the walls should be 2 inches (5 cm). Again, using rough sawmill lumber gives extra thickness, because it isn't planed. These planks should be spiked using at least 20-penny spikes. (I like to predrill the holes in the planks with a drill bit that's smaller in diameter than the spike to avoid splitting and the resulting weakness.) And as with any livestock building, the planks should be nailed on the *inside* of the structure. Yes, planks on the outside look nicer — but they aren't nearly as strong. A horse can actually kick them off the posts, nails and all! But when he kicks a plank that's nailed inside, it just goes "bang" and remains in place.

Where economy is of prime importance, consider using 6-inch-diameter (15 cm) logs harvested from your woods or purchased inexpensively from the U.S. Forest Service or a local supplier of firewood. Flatten these logs on both the top and bottom to use as wall partitions in the stall. Another alternative is using thick slabs from a sawmill. (Slabs are the rounded, log-siding-like pieces that are cut off a log before it is sawed into lumber. Depending on the mill and the use the lumber will be put to, these slabs can be quite thick and even, making for a great inexpensive building material.) Both of these choices can result in a very attractive horse stall.

Quick Tip

Use pressure-treated planks for the bottom two rows of planking, which will save you a considerable amount of money and energy on later replacement of rotted planks. However, I don't recommend using treated planks for the rest of the stall walls. They contain toxic chemicals, and I don't feel they're safe to use where horses may be in constant contact with them, or possibly lick them.

You need to plank the walls of your stall or run-in shelter up at least 4 feet (1.2 m) high inside a building or stall — I like to plank the walls even higher if I'm housing a stallion, a young horse, or a very active animal, because the kicking may well be higher. (See also page 23 for information on building and planking a stall in an enclosed shelter.)

Step 1. Begin planking at the bottom, very close to the floor. If you begin at the top, there's often a large crack left at the floor level, and a horse can get a hoof through a surprisingly narrow opening when rolling and become severely injured. Use a level before nailing or screwing each plank, and adjust each plank slightly to compensate for any irregularities.

In very hot climates, you may want to leave an opening between planks for additional ventilation. Don't let the gap exceed 1 inch (2.5 cm), though, or your horse may begin to chew on the planks. And again, a horse (especially a foal or small pony) can kick his hoof through a very small opening.

Step 2. Plank up at least 4 feet (1.2 m). If you're planking one stall that adjoins another, it's wise to plank all the way up to the ceiling to prevent even friendly horses from biting and kicking at each other in play. This avoids many injuries, and it makes the stall partition last years longer.

Step 3. Build up the sides and back from the top of the planking to the roofline using more 2 x 6 planks, 1-inch (2.5 cm) board-and-batten construction, siding, or sturdy plywood.

Building a Simple Run-In Shelter

Where only one or two friendly horses need to be housed and you don't have an existing structure that lends itself to proper renovation, often the best way to go is to build a simple run-in shed from scratch.

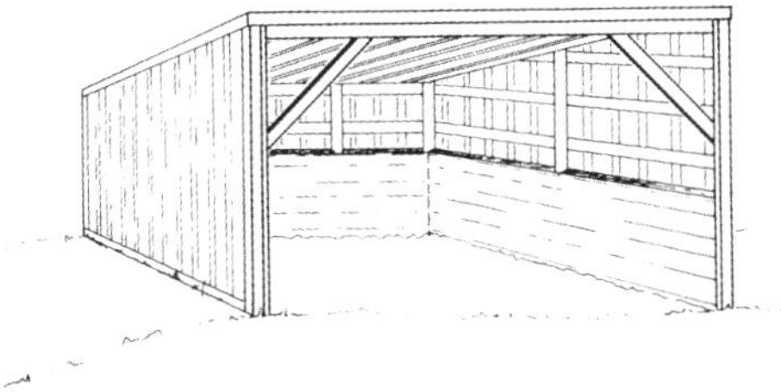

This three-sided shelter is set on a site that drains well. It's planked to 4 feet (1.2 m) on the interior and finished with metal sheeting on the exterior, supported by interior 2 x 6 braces.

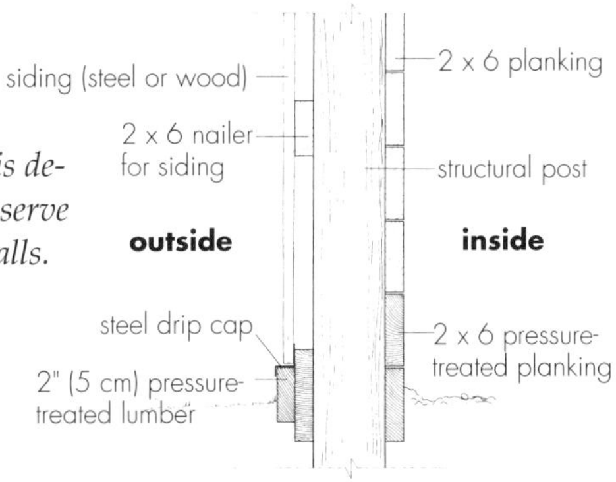

The wall of the three-sided shed is designed to prevent injury and preserve the building's foundation and walls.

This three-sided shelter typifies pole-building construction. It uses planking on all sides, with pressure-treated planking at the base and a corrugated metal roof.

This type of shelter has an added advantage of not requiring daily bedding and cleaning the way a stall does. This is not to say that the run-in shed should be dirty or wet; still, its very airiness — especially if it's located on a slight hill for good natural drainage — precludes wetness. If you can muck out the shed fairly well once a week and thoroughly clean it twice a year, it will usually remain quite clean.

Figuring Dimensions

A shed for one or two horses can be quite small — about 12' x 15' to 20' (3.7 x 4.6 to 6.1 m) is adequate. This size allows an animal to move about to escape blowing rain or snow, but it's still small enough to be economical for you and easy to clean. It's best if the shed can be freestanding in the pasture; this allows the horse to take full advantage of its protection no matter where the sun or wind is coming from.

The back wall — often the lowest wall — should be at least 7 feet (2.1 m) high to avoid head injuries to the horse. Depending on the climate and expected snow load, the roof can slope somewhat or be steep enough to allow heavy rain to run off instantly and heavy snow to slide off before damage is done. The front of the shed is generally about 10 feet (3 m) high, which also allows you to clean it (at least in part) with a tractor and a manure bucket or a smaller skid loader.

Choosing the Right Materials

While many run-in sheds are constructed solely out of sheet metal, this is not the best or safest choice for horses. I've seen many, many injuries result from horses kicking through the lightweight metal, kicking a foot under the sheet metal, or ripping open a body part on a torn, projecting piece of metal. Back the sheet metal with rough-sawed 1-inch (2.5 cm) lumber, which is thicker — that is, stronger — than planed lumberyard lumber or ¾-inch (1.9 cm) plywood. Even so, it's a good idea to line the inside of the shed to 4 feet (1.2 m) up with 2-inch (5 cm) planks for additional safety against kicks.

Give careful consideration to the shed's walls. Besides the common sheet metal, other materials can be used for both economy

Innovative Ideas

A friend on a very limited income built her shed's sides with logs from her woods flattened on one side. The 6-inch (15 cm) logs were strong, long lasting, and made her horse shed look like a rustic log cabin nestled in the trees. Another innovative friend bought some used steel pipe and ⅛-inch (3 mm) steel plate inexpensively at an auction and, being a talented welder, built his run-in sheds out of these materials. They were attractive, safe for the horses, and very long lasting.

and horse safety. One of my barns was sided with 4' x 8' (1.2 x 2.4 m) sheets of grooved cedar siding screwed onto 2 x 6 sawmill lumber placed every 2 feet (60 cm) up the barn wall's posts. At ½ inch (13 mm) thick, backed by the 2 x 6 planks, I had to repair only one "kick hole" in 12 years — and no damage to the horse!

If you simply must use unbacked sheet metal to side your run-in shed, keep these walls outside the horse paddock or pasture. Locating such a shelter in a corner eliminates two sides that need backing to be strong enough to resist the kicks and bumps a horse is sure to provide. But putting the shelter in the corner of the pasture also eliminates two sheltering sides for the horse should the wind blow from an unusual direction or snow be severe. At any rate, the inside of the shelter must be planked; otherwise, your horse will be at great risk for a serious injury.

Framing Up

The frame for the run-in shed is simple. Set used telephone or electric poles 3 feet (90 cm) into the ground, and preferably also into a concrete base, no more than 12 feet (3.7 m) apart. Onto this base, spike or bolt 2 x 6 or greater framework for the walls and roof support. Set 2 x 6 roofing rafters no farther apart than 24 inches (60 cm). You can economize by using 6-inch (15 cm) logs, flattened on two sides, for rafters. Even when using sheet-metal roofing, it's best to use at least 1-inch (2.5 cm) rough-sawed lumber — spaced 2 inches (5 cm) apart if you have to economize — for the rafters. This strengthens the building against strong winds and also keeps the roof on a lot longer.

The Roof

> **Quick Tip**
>
> When using sheet-metal roofing, use screws with gaskets made for this purpose, because they hold much better than nails and prevent leaks.

You can use asphalt shingles or even rolled roofing, but I really recommend sheet-metal roofing. It lasts longer and, more important, it's fire resistant. (Forest and grass fires can occur almost anywhere; their flying firebrands destroy many buildings after setting fire to a shingled roof.) The roof should extend at least 2 feet (60 cm) beyond the back of the building and 3 feet (90 cm) in front, for added protection from the elements and moisture for both the horse and the building itself.

Making Provisions for Food and Water

While some folks feed and water inside their run-in shelters, I prefer to do so outside at some distance from the shed. First of all, nearly every water tank or self-waterer will at some time overflow, causing wet ground. Some horses even like to play in the water, splashing it about or letting it overflow by holding the valve open with their nose . . . just for fun. You don't want this wet inside the shelter. (Also, it's hard to keep the horse from depositing manure in an indoor waterer or tank.)

Feed presents different problems, especially if two horse "buddies" share the same shelter. Even between buddies, fighting at feeding time is common, and you don't want this happening inside the shed. It causes damage to a building and possible injury to a horse. Roughhousing outside, though, is usually of no consequence.

Another reason to locate the feed trough or box outside is for your own safety. All horses are impatient at feeding time, prone to shoving, biting, or kicking in annoyance at not getting their feed *right now.* If you're carrying a bucket of feed and hay bales through the pasture and into the limited, enclosed space of a run-in shed, you have the risk of injury. This is greatly multiplied by every other horse in the pasture sharing the shed at feeding time, no matter how gentle or kind he may be. It is much better to be able to feed the horse(s) from outside the shed and pasture. Locating feed boxes just over the fence provides ideal safety and time efficiency.

Designing the Layout for a Small Barn Renovation

There are many types and styles of economical small horse barns, but all of the good ones have a few things in common:

- They are safe for a horse to live in.
- They are built to last many, many years.
- They are safe and easy to work in.

Sounds simple, doesn't it? And it can be, even if you're planning on renovating an existing structure to make it fit to be called a horse barn. However, it's important to take a good, critical look at the structure and its location before becoming too deeply involved in the project. If it's an older building, you may even want to have a professional contractor examine the integrity of its frame and foundation. Reviewing the points discussed on page 4 will help you make sure that the structure has the potential to be a good, safe shelter for horses before you get started.

Stall Dimensions

Although horses can certainly be housed in tie stalls (a narrow stall in which the horse is kept tied) — and have been since biblical times — the modern horse often gets less-than-optimum daily exercise. Thus, it's best to house him in a roomy box stall where he can lie down comfortably, walk around at will, and even play a bit when the mood strikes him. Plan for a 10' x 12' (3 x 3.7 m) stall, which is

Free Range

It's crucial to your animal's health and well-being that his barn provide access to a good pasture or large paddock for daily exercise and grazing. Horses are open-space creatures, and to lock one up in a stall, day in and day out, is cruel. It nearly always causes mental and physical problems in equines, no matter how expensive the barn or how good the care. Bad habits like cribbing, wind sucking, weaving, and pawing are nearly always caused by boredom.

Sample Designs for a Small Barn Renovation

A simple one-horse, one-story barn with a drive-through aisle

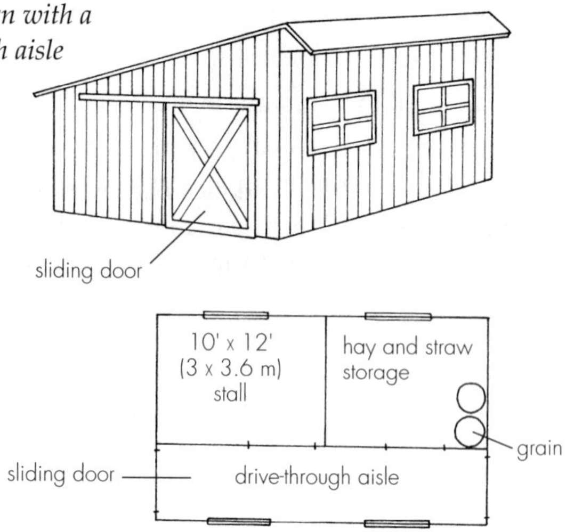

A simple two-story barn with a haymow

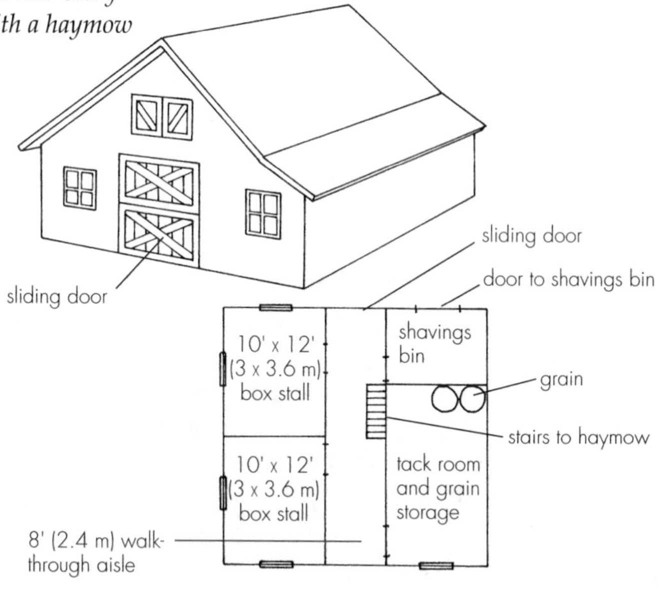

commonly accepted by a single horse and can accommodate a mare and foal as well. A minimum of 7 feet (2.1 m) is required for headroom; 8 feet (2.4 m) is even better. Any less is asking for head and neck injuries. (For details on building the stall, which takes place after the rest of the renovations have been completed, see page 23.)

The Frame and Walls

Like a renovated run-in shelter, the frame of an existing structure being remodeled for use as a small horse barn needs to be reinforced with posts set in concrete and joined to the existing members (see page 6). Interior walls that will become part of the horse's stall or that the horse might come in contact with need to be renovated accordingly (see page 23 for details on building the stall). Exterior walls that the horse will *not* come in contact with simply need to be sturdy and in good shape.

Insulation in a small horse barn is nice but not necessary. Horses do fine in an uninsulated barn, no matter how cold the interior gets, provided they are well bedded, dry, and not in drafts. In fact, more horses are made sick by barns that are too warm than too cold. For this reason, a horse barn should not be heated.

Benefits and Drawbacks of Insulation

Benefits

- The barn will be warmer in the winter and cooler during the summer.
- The barn will be less drafty.
- Water won't drip as frost melts or as cooling and warming temperatures cause heavy condensation.

Drawbacks

- The cost is greater, as is the labor involved in construction.
- Insulation must always be protected from horses, because they will sometimes eat it, then become sick, even to the point of death.
- The barn sometimes becomes too "close," requiring extra ventilation to prevent respiratory problems.
- Insulation provides nesting for rodents.

Installing a Sliding Door

Sliding doors can be made of 1-inch (2.5 m) lumber, framed and reinforced with the same. Using 4' x 8' (1.2 x 2.4 m) sheets of ½-inch (13 mm) barn siding framed with 1-inch lumber also works well. Be sure to install the rollers and track correctly so that the door slides easily and doesn't drag. If the opening is 10 feet (3 m) wide, it's best to use two 5½-foot (1.7 m) doors that can slide either way, for ease of handling and less individual door weight.

I like to drive a 2-foot (60 cm) piece of 1-inch (2.5 cm) steel pipe a foot and a half (45.7 cm) into the ground on each side of the door opening to brace the door and keep it from swinging outward, which can happen during a strong wind or even when a loose horse shoves it. A bottom roller is meant to prevent this, but the pipe is extra protection.

Even though your barn has a sliding door, you should still have at least one walk-in door to lead your horse in and out. This is handy since a sliding door requires some effort to open, and there's a tendency to not open it far enough when leading a horse through. I've seen several doors wrecked and owners and horses injured when a horse caught the door with his hip while hurrying in or out. A walk-in door is also an additional safety measure in case of fire.

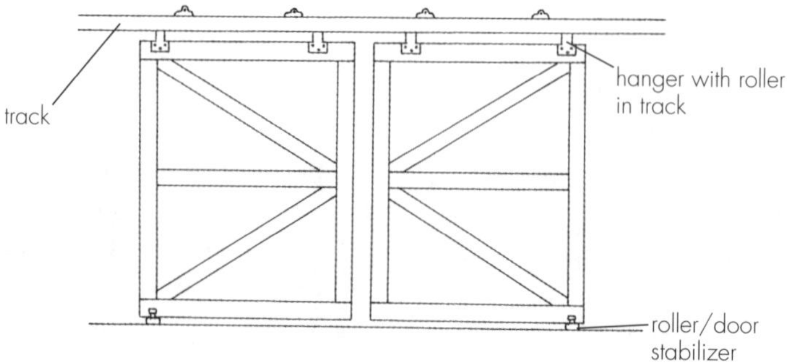

A sliding door with a long track allows easy, drive-through delivery of heavy materials.

Aisles and Walkways

If you're working on a relatively large structure, consider adding a wide inside aisle to your plans. Such an aisle is very handy, especially during winter and wet months. Eight feet (2.4 m) in width is adequate, but 10 feet (3 m) is better and will allow a small tractor and manure spreader to be driven through the barn for quick, easy daily cleaning — a sliding door at each end of the aisle makes for simple drive-through access. An aisle also allows for easy delivery of heavy materials, such as fill dirt for stalls, feed, and bedding. The indoor aisle is also a good place to locate a cross tie where a horse can be groomed and sprayed for flies, have his feet cleaned and trimmed, and have minor veterinary procedures performed.

Doors

Doors should be a minimum of 3 feet (90 cm) wide and should either slide fully open or be able to be swung easily open on very strong hinges. All stall doors should swing into the aisle, because any bedding shoved up to the door will restrict the opening if the door opens into the stall. Then, if the horse is impatiently waiting to go outside, he may rush into the half-open door and become trapped in the narrow space. The more he fights to free himself, the more severely he can be injured.

Flooring

As for flooring, my opinion is that the less cement in a horse barn, the better. Concrete is especially slippery to shod horses and causes many falls. It is also, of course, very hard and unyielding, causing injuries to feet and joints. And it's cold, especially when used in the stalls where a horse would like to lie down comfortably to rest. Yes, cement is easy to keep clean, hose down, and so on, but I strongly feel that its minuses far outweigh its pluses. The nicest barn I've ever worked in had used-brick aisles and stall floors made of dry, hard-packed clay. The brick made a less slippery hard surface that was still easy to sweep and hose down — besides looking great. Sometimes used brick is available by the truckload for free or very cheaply at demolition sites. It pays to ask around! Can't afford or can't find used brick? Plain hard-packed dirt, especially if it contains some clay, is fine. You won't be able to hose it down, but you can rake and sweep it clean.

Windows

Every stall should have a decent-sized window — placed high up. This lets the horse look out but doesn't allow him to lean into the protected screen or kick it in play. A window at least 18" x 24" (45 x 60 cm) is good. It should be able to be slid open for ventilation, be screened against biting insects, and be covered with hardware cloth to protect the screen from the horse's inquisitive nose. I prefer using acrylic plastic instead of glass for horse windows, just in case there's breakage. Plastic doesn't shatter, so it's much safer.

Water

Before you begin any renovations, consider running a water line. Having water in a barn is a great labor saver and will result in the horses having more access to fresh water: Obviously, it's easier and quicker to water the animals where water is readily accessible. Where freezing is a problem, an underground (below the frost line) water line, using economical flexible black poly, can be run from the well or the nearest water source. If you attach this to a frost-free hydrant, which drains completely after use, even a barn that doesn't remain above freezing during the winter can have plenty of water handy. (Take special care to locate the hydrant in a very protected spot so that it can't be damaged by equipment or horses. One of the worst horse accidents I've ever seen happened when a playful horse caught his halter on a frost-free hydrant and tore his head and throat terribly, exposing 6 inches of the jugular vein. Luckily, with immediate veterinary treatment he recovered fully.)

Having water in the barn provides a bonus: fire-fighting ability that just may save your horses and barn in the event of a fire. Always keep at least a 60-foot (18 m) length of hose coiled on the wall, dry and unfrozen. The hose that I designate the "fire hose" isn't used during the winter, just to be sure it doesn't become clogged with ice. My "horse-watering hose" is separate.

Hay, Bedding, and Feed Storage

While planning the layout of your barn, you need to consider just where your hay/straw, bedding, and grain will be stored. While in some drier climates it's possible to store hay bales outdoors, this practice is wasteful. Not only will some moisture find its way into and between the bales, causing probable mold and spoilage, but even minute condensation can create dusty, moldy hay as well. This mold and dust can cause many health problems in horses, frequently colic and heaves, an asthmalike condition. And no, covering the hay with plastic tarps won't protect it from condensation, even though rain and snow may not penetrate. Fresh hay gives off some heat and moisture, and when it's tightly covered and standing out in the sun, plastic tarps cause it to sweat. This moisture appears all through the hay and will soon produce that dusty, moldy, musty smell.

Having a two-story barn is the ideal situation, but if the structure you're working with doesn't lend itself to this, simply add another box stall specifically for hay and straw, shavings, and grain is as easy solution. If you need more space than this, you can add a "back porch" or shed room off its side, which can double your storage capability quite inexpensively. A simple addition can be made by extending the sloping roof down to a convenient height — 6 feet (1.8 m) or so — and walling in from there.

Benefits and Drawbacks of Using a Hayloft

Benefits

- Lofts make it easy to store and retrieve hay bales.
- No horse can get loose and have access to the hay, which causes much destruction.
- The barn will be warmer during the winter and cooler in summer due to the insulating properties of the bales.
- You'll need less ground area than if you included storage on a single floor.

Drawbacks

- Two-story construction is more costly, in both money and labor.
- To support the weight of the hay bales, strong bracing must be provided underneath, with at least 2 x 8 floor joists or log joists required.
- The loft must be floored. This can be done with rough-sawed 1-inch (2.5 cm) lumber, but it does cost more and takes a few days longer to construct.

The feed room is one place I heartily *approve* of concrete in the horse barn. Not only does the cement keep the bottom tier of hay or straw dry, dirt-free, and mold-free, but it also prevents our little rodent friends from gaining easy access to a free lunch. However, for an added measure of protection, do make sure that the floor has a gentle slope to it, and that it slopes toward a drain.

Do yourself a great favor: Close in the front of the feed storage room and add a sturdy 4-foot (1.2 m) access door. This will almost positively ensure that a loose horse can't get into the feed and gorge himself. A tight-fitting lid on each feed drum will help, but a door is nearly foolproof. A horse can usually get away with eating all the hay he wants, but grain is another story. Pigging out on grain will often cause colic, severe digestive upsets, and founder.

Some feed areas are *too* tight. This causes the hay to sweat, much as if it were under a plastic tarp. Add extra ventilation like a roof ventilator or a small outside window and a window opening into the barn aisle (well screened, of course).

Storing Feed and Bedding

Stack straw separately from hay, while providing convenient access to both. If you use wood shavings as bedding, make sure you keep them separate from the hay, which, if ingested, would result in coughing horses. It's best to build a simple plywood bin to store the shavings; a light tarp cover for the bin is an additional help. Locating this bin with an outside access door is a help. That way you can back your truck or trailer up to the open door and simply shovel in the shavings. Any labor saver is great, in my book!

Grain can be stored either in a separate enclosed wooden bin, built on the cement, or in clean, nontoxic 55-gallon (208 l) covered plastic or steel drums. Be sure you know what was in them before you got them. Drums that contained detergents, cooking oil, or other such benign substances are fine. But beware of fuel, chemical, or fertilizer drums. No matter how well you clean them out, your horse can still be exposed to toxic fumes or material. You don't want a "glow-in-the-dark" mount!

A Tack and Tool Area

A single-horse barn can get along quite well without a separate tack and tool area if you use an out-of-the-way corner or even a separate shed to house such items as saddles, bridles, blankets, pitchforks, shovels, and wheelbarrows. But when there are two or three horses in the family, it's best, if at all possible, to include a small tack room in your barn plans.

This room can be small, depending on the amount of tack and tools that will need housing. I've seen a tack room of only 6' x 10' (1.8 x 3 m) function very well, with stacked saddle racks; bridles neatly hung on the wall; a box for winter blankets, summer sheets, leg wraps, and the like; a small cabinet for veterinary and grooming supplies; and tools hung well out of stumbling reach.

Like the feed room, the best tack room is an enclosed, solid-floored room with a sturdy door to keep out rodents, dogs that might chew leather, and dust — not to mention the infamous loose horse. And I've had them in my tack rooms even with a door!

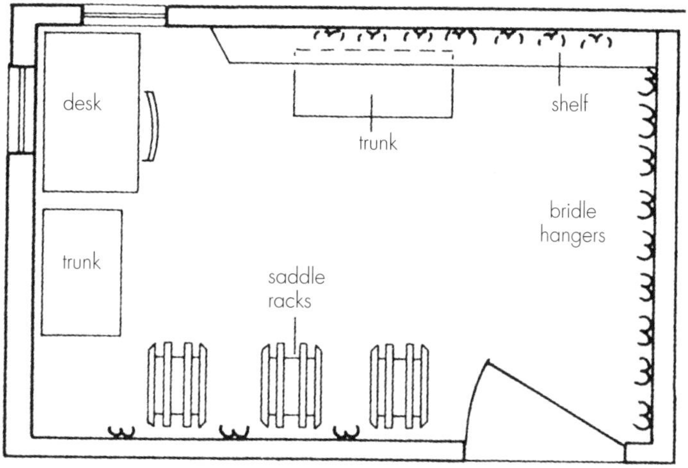

A simple, functional tack room makes good use of shelves, racks, and hangers to keep gear off the floor and out of the way,.

I like to insulate the tack room then, finish it with inexpensive paneling, since it is a place of frequent work and the insulation keeps it cooler in the summer, warmer in the winter. A properly installed small heater, used only when work is in progress, provides safe comfort for the worker.

Any horse supply catalog or tack shop can be the source of ideas for saddle racks, bridle hooks, tool hangers, and other handy items for the tack room. Now, I don't necessarily say "buy." I'm a real do-it-yourselfer; I've gathered many ideas in catalogs or shops that I then rework at home from readily available materials. Not only does this save a lot of money, but it provides an outlet for artistic talents as well, making the tack room truly one of a kind.

Fencing around the Barn

Finally, always build extra-strong, extra-safe fences near the barn, because there is concentrated activity in that area. Plank, rail, pipe, or some other material that will safely withstand shoving, lunging, and kicking is necessary. During the design stage, plan how your fences will relate to the barn. Some folks prefer to have a wide lane running from the barn door out to a pasture, or separate paddocks. The less contact horses have with the outside walls of their barn, the better. Sooner or later they will do some damage to the siding, whether it be dings and scrapes, chewing, or kick holes.

Fire Extinguishers: An Absolute Necessity

A *must* for the horse barn is a fire extinguisher, kept right beside the door. In fact, one placed next to *every* door is even better. You never know when a fire might start — from electrical failure, wiring damage, lightning, careless smokers, and much more — and it seems as if they always start at night. Having extinguishers handy at the door provides instant help and prevents the panic of having to fight through flames to reach them.

Likewise, at least two fire/smoke alarms are necessary in every barn. Locate them some distance from each other, one near the tack/feed room area, the other at the opposite end of the building. If one fails, the other may save your horses.

Roofing Your Horse Barn

Several roofing materials are available, but the best system I've seen used is plywood covered with corrugated sheet metal. The sheet metal is long lasting, resists hail and wind damage when properly screwed down, seldom leaks, and — its greatest advantage — protects against grass and forest fires. Many barns and other structures are lost to fires started by burning firebrands carried in the wind, and this alone is reason enough to use sheet metal on your barn roof.

But used alone, and on an enclosed barn, sheet metal can cause condensation problems such as frosting up on cold nights. This frost forms on the underside of the roof, to later thaw and drip down on the horses. In most cases you can prevent this by using ¾-inch (1.9 cm) plywood underneath the sheet metal. This generally prevents heavy condensation problems, plus it will greatly strengthen the roof against wind and heavy snow loads.

In northern areas, where heavy snows are common, be sure your roof slopes steeply enough to quickly shed falling snow. A snow-covered roof may be picturesque, but that pretty white stuff weighs tons — literally — and can weaken or even cave in a barn roof if it stays up there long enough.

Building a Stall

Individual stalls themselves are constructed after the barn is "dried in" — when the shell is built.

Step 1: Reinforce the frame. Set sturdy, 6- to 8-inch-diameter (15–20 cm) poles or squared timbers 3 feet (90 cm) into the ground in a concrete base at each of the inside stall corners. Make sure that these poles are strong and secure, because they form the framework for the stall. (And if you're building a large stall, add additional poles as necessary so that no pole is more than 12 feet, or 3.7 m, from another.)

Step 2: Plank up the walls. Once the corner poles have been set, you're ready to plank the stall. Nail or screw 2 x 6 (or greater) planking to the corner posts to form the walls and front of each stall, up to a height of at least 4 feet (1.2 m). Remember to secure the planks on the inside, or "horse side," of the stall. This provides greater strength and safety than if the planking is fastened to the outside of the poles.

Step 3: Cover the back. Cover the top portion of the back wall with ¾-inch (1.9 cm) plywood, well supported by underlaying 2 x 6s every 2 feet (60 cm) for strength.

Step 4: Cover the side walls. What you use to cover the upper portion of the side walls of the stall depends on where they fall in the layout of the barn and your own preferences. The top portion of the side walls can be covered with plywood, like the back wall, or you can use a wire mesh or a welded-pipe grate, like the front wall.

Step 5: Cover the front wall. The top portions of the front wall, including the door, can be covered by a heavy-gauge wire mesh or, better yet, a welded-pipe grate. Construct the pipe grate out of used ¾- or 1-inch (1.9 to 2.5 cm) galvanized steel pipe, welded onto a 2-inch (5 cm) channel iron the full length of the area to be covered. The pipes should be no farther apart than 4 inches (10 cm) — closer if the horse to be housed has small feet. I've found it very helpful to have one of the pipes set in a hole at the place you plan to feed, rather than welded. This lets you feed without going into the stall by simply sliding the pipe up through the top hole and dumping the feed into the box or pail. Aside from being a great labor saver, it's also safer for you, since accidents sometimes happen at feeding time when an impatient horse rushes to eat.

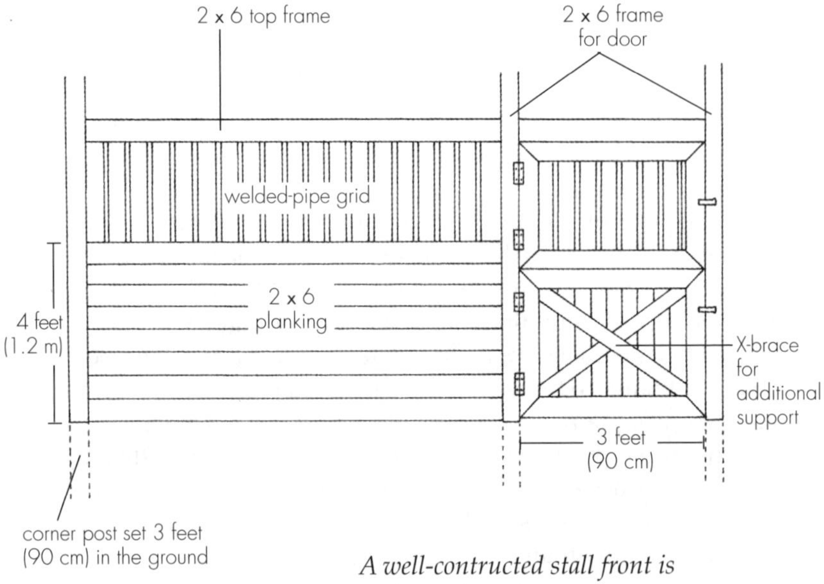

A well-contructed stall front is both sturdy and attractive.

Making Do with Stock Panels

Many people use livestock panels for the top of their stall fronts. These are inexpensive, very strong, and easily purchased through any farm supply store. They come in 16-foot (4.9 m) lengths and are usually just under 48 inches (1.2 m) high. Shorter "hog panels" are also available should your ceiling height be less or if you decide to plank up higher. The mesh in these panels is about 6" x 8" (15 x 20 cm) and is spaced closer near the bottom quarter of the panel. These livestock panels can easily be stapled to the uprights with long livestock fence staples, and they can be cut down to size using bolt cutters.

One caution about using stock panels in this way: A frisky horse can get his foot through the opening in the mesh. A staid old-timer is probably fine in a stall with a stock panel front, but it's best to line this front with 2" x 4" (5 x 10 cm) welded wire for extra safety. The cost is minimal and may prevent unhappy accidents in the future. Chain link is another alternative, as is the 2" x 4" wire used alone. Either can work, but both will have to be replaced at some point, because with time they'll both sag from being rubbed on, lunged against, and kicked.

Step 6: Frame the stall door. The stall door can be framed with 2 x 6 lumber reinforced with a 2" x 4" (5 x 10 cm) X-shaped brace. Tongue-and-groove 2 x 6s or ¾-inch (1.9 cm) plywood can be used to build the door, again nailed or screwed on the "horse side," where the pressure will be. Again, this door should be a minimum of 3 feet (90 cm) wide.

A Dutch door (which has a top that opens separately from the bottom) is very nice for a gentle horse, allowing him to hang his head over the lower door to watch goings-on down the aisle. But an aggressive or very frisky horse should be housed in a stall with a floor-to-ceiling door with a small grate at the top for ventilation and also to look out. No horse should be locked up like a prisoner in a stall with no view outdoors and into the aisle. (*Note:* If you're using a Dutch door, arrange a way to fasten its top half open flat against the stall grate. Otherwise it can blow shut in the horse's face. Even worse, he will also quickly learn to play with his new "toy," shoving the door top open again and again just to hear it bang!)

Step 7: Bolt the door to the stall. Whichever style of door you choose, use very strong hinges, bolted to the door for extra holding power. Likewise, the latch should very heavy duty — and horse-proof! A simple hook and eye is not good enough. Nearly every horse quickly learns to unhook his stall door and go exploring. A multiple-function hook — which requires a latch to be lifted before the hook can be lifted — or a heavy barrel bolt latch works much more satisfactorily.

Providing Food and Water in the Barn

All horse stalls need to be equipped with the means to provide food and water to the animal occupant. Because horses are such large and playful creatures, you can't just plop down a water bowl and a food dish like you can for Rover and Kitty and say, "There you go." Horses take a bit more planning.

Water Buckets and Holders

Each stall should have provision for water. This is most easily and economically furnished by hanging a wide, indestructible 5-gallon (19 l) water pail — either rubber or heavy poly — in the stall. While you can just suspend the water bucket from a heavyweight snap around an eyebolt through the stall wall, I like using a welded bracket as well. The bracket is simply two rings welded onto two flat ¼-inch (6 mm) steel uprights, with a bar or two across the bottom to support the bucket. The flat, upright bracket is bolted to the wall at a convenient height by means of two holes in the wall side of

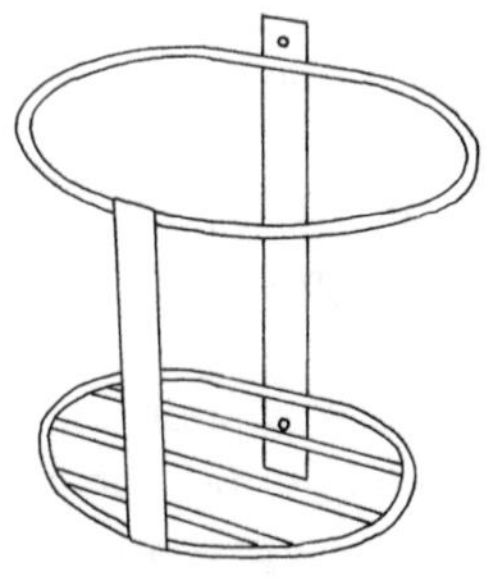

A water-bucket holder

the stall. This bracket can be purchased through most horse supply catalogs or made by a handy do-it-yourselfer with welding skills.

The reason I like the bracket is that it discourages a horse from playing with and dumping a partially full bucket — and they *will* play with the water bucket! The bracket will also help the water pail last longer, because the weight of the water will pull on the bucket constantly, reducing its life.

Treated right, rubber pails will last almost indefinitely. I have some that are 20 years old. They look a bit worse for the wear, but they still hold water!

Salt Blocks

Each stall should have a salt-block holder so that salt is available to the horse whenever he needs it. A stall block is smaller than a field block — usually about 2 to 4 pounds (0.9–1.8 kg) — and if it's just placed in a tub, the "helpful" horse will soon dump it out onto the floor, where it will quickly disappear.

Instead, fasten a salt-block holder to the wall so you can simply slip in the block of salt from the top. Result: No waste, no mess.

Hay Feeder

There is quite a debate among horse owners regarding the "right" way to feed horses hay. I feel that the best way is simply to place it on the floor in a clean, dry corner of the stall. This allows the horse to eat in a natural position, and there is absolutely nothing else in the stall for him to get injured on.

But many folks feel that a separate hay feeder on the wall is better. This welded steel bracket with slanted bars on the front allows you to drop in "sandwiches" of hay from the top. Usually a tray sits below to catch the fines, or leaves, preventing waste.

> **Quick Tip**
>
> *Never* feed a horse on the floor of a dirty stall. If he is exposed to filth, you may well have a recurring internal parasite problem. If you don't want to use an elevated hay feeder, be sure that the floor of the stall is clean before leaving hay there for your horse to feed on.

A wall-mounted hay feeder is a pretty good setup for most horses. But if a horse is very playful, he can find ways to injure himself on most steel feeders at one time or another. To prevent those sorts of accidents, if you choose a steel feeder, be sure to pick one that has rounded, smooth edges, is very solid, and has closely spaced bars. Then locate it by itself on a wall away from the water bucket.

A wall-mounted hay feeder

An inexpensive alternative to the hay feeder is a hay net — a poly or nylon net bag that you fill with hay and hang on an eyebolt, just above your horse's head level. This works quite well and is pretty safe (although some creative horses manage to get a foot hung in them). It is awkward to use, however, taking more time to fill than a feeder. In addition, it allows waste of the fines, because they trickle to the ground while the horse eats.

Grain Feeder

It's necessary to supply a separate feeder for grain. This is most handily mounted on the front wall of the stall, where the grain can be poured into it through either a "door" in the pipe stall grate or a "window" cut into a stock panel or wire-mesh grate, framed by welded steel.

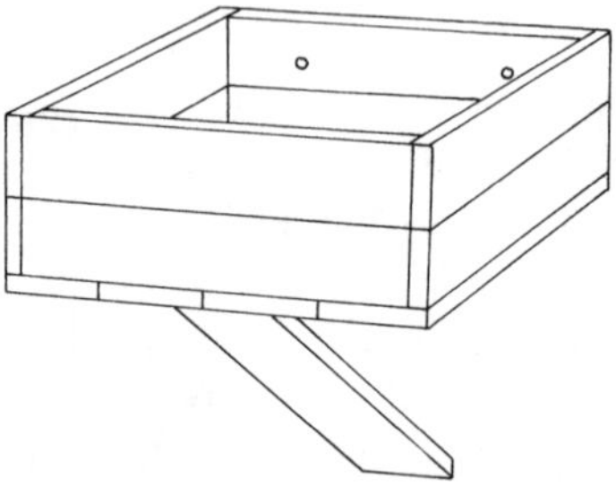

A wooden feed box

A grain feeder can be a bucket — held by the same type of bracket as the water pail — or a very sturdy wooden box made out of 2 x 6 lumber. The dimensions of such a box can vary, but it needs to be at least a foot (30 cm) deep or the horse will toss out and waste feed as he eats. For extra strength, instead of nails use screws long enough to get a good bite into the wood. Bolt this box securely on the stall front at or a bit below 4 feet (1.2 m) — depending on the size of the horse. An extra bracing bracket of 2 x 4 lumber or a 4-inch (10 cm) pole cut at an angle will help keep the feeder in place for years.

It's a good idea to put the grain feeder in the opposite front corner from the water pail. Having them side by side, as some folks do, often means having bits of grain in the drinking pail — which quickly ferment.

Finishes to Preserve Your Horse Shelter

After the barn has been constructed, some work still needs to be done. For the barn to last — not to mention remain attractive — it requires an exterior wood finish such as stain, wood preserver, or paint, as well as a nontoxic interior finish both inside and outside the stalls. Of course, if you used sheet-metal siding for the exterior walls, the outdoor finishing will be minimal. But any type of wood construction — whether siding, log, or board and batten — must be effectively protected to withstand the elements and provide you with many years worth of service as a strong, stable shelter for your horse.

The Pros and Cons of Paint

Paint comes to mind first off, but it's not always a good choice. Paint forms a shell over the wood, and no matter how good the primer and paint are, sooner or later some of the finish will begin to peel and flake. (And if you made the mistake of using a cheap paint or primer — or worse yet, no primer — this peeling will begin all the sooner!)

Never use paint left over from a long-ago job, which you may have discovered in a shed somewhere. Old paint is often lead based, and this is toxic not only to humans but to horses as well — and they are much more likely to ingest some of it.

Tips for Using Wood Stains

I feel that stains are a better choice for the finish of horse barns. If you've used very nice wood, such as log construction, you won't want to cover it up with a solid stain that looks like paint. Instead, use a clear stain, which allows grain and texture to show and makes your barn a work of art. If you've used wood siding or 4' x 8' (1.2 x 2.4 m) panels, choose either solid or opaque stain. Solid stain comes in a wide variety of colors and looks like paint on the wall, but the stain *penetrates* the wood, locking into it, so it won't peel or flake off over time. It may fade a bit, requiring another coat in a few years, but it is easily reapplied without scraping and sanding.

Stains are either oil or latex based. Latex stains have the advantage of easy cleanup, but oil-based stains offer better protection to the wood. Stain can be applied to the barn wall with a paint sprayer run by an air compressor, a paintbrush, or a roller. The choice will depend on the type of wood you've used as well as your personal preference.

Finishes for the Interior

The inside of the barn can receive the same treatment — perhaps in a lighter shade or color than the exterior, since a lighter color reflects more light, making the barn seem less dingy. Be absolutely sure that you use only a nontoxic stain or finish on the interior of the stall; there's always the possibility that a horse may chew or lick it. I've been quite pleased with boiled linseed oil, painted on while still quite hot, as an interior finish for stalls.

If your barn has ceilings, a coat of white solid stain works well and seems to discourage flies a bit. Better yet, give that ceiling two or more coats and you'll be able to scrub off any dirt or fly specks with ease.

Always do any interior painting/staining while the horses are outside, because the fumes can cause respiratory problems. (And of course provide adequate ventilation — open windows and doors and use a fan — for your *own* safety and comfort as you work.)

Tips for Electrifying the Small Horse Barn

Electricity is a great help in a barn. But unless you're an accomplished do-it-yourselfer with experience in electrical hookups, leave this job to an electrician. Many fires have been caused in barns because of faulty wiring or improperly installed boxes or light fixtures.

Generally, one light fixture in the aisle placed every 12 feet (3.7 m) along with one in the center of each stall works quite nicely. Add a light in your mow, feed room, and tack area for convenience. And unlike houses, a plug-in outlet every 6 to 8 feet (1.8– 2.4 m) isn't a necessity — but it is very handy to have one outlet on the outside of each stall, in the tack area, and where you have your cross ties and do your grooming.

Of utmost importance, be sure to protect all electric wiring, outlets, and fixtures from inquisitive or rambunctious horses:

- Purchase heavy metal cages to protect lightbulbs from tossing horse heads.
- Protect any exposed wiring with an aluminum conduit. Better still, run the wiring inside a wall or the ceiling.
- Locate any switches or outlets way out of horse reach. If you don't, they will be nibbled and crunched — and the horse may get a bad shock in the process.
- Be absolutely sure that your wiring has been sufficiently grounded to prevent shock and accidents.
- If you ever have to use an extension cord, use a heavy-duty one placed well out of reach of equines, and do away with it as soon as you can. This may prevent an accident or fire.

Other Storey Books You May Enjoy

Building Small Barns, Sheds & Shelters, by Monte Burch. Covers tools, materials, foundations, framing, sheathing, wiring, plumbing, and finish work for barns, sheds, garages, and more. Paperback. 248 pages. ISBN 0-88266-245-7.

Getting Your First Horse, by Judith Dutson. This book will answer hundreds of questions prospective horse owners have, such as: What do I need to know to choose a horse? What will my horse need for housing / feed / exercise? How much will all this cost? Important medical terms, horse-handling techniques, a glossary, and references are included. Paperback. 176 pages. ISBN 1-58017-078-1.

Horsekeeping on a Small Acreage: Facilities Design and Management, by Cherry Hill. The essential guide for designing safe and functional facilities whether you have one acre or hundreds. Paperback. 192 pages. ISBN 0-88266-596-0.

How to Build Small Barns and Outbuildings, by Monte Burch. Contains complete plans for 20 projects, including an add-on garage, workshop, roadside stand, equipment shed, greenhouse, and four types of barn: all-purpose, pole, horse, and milk. Paperback. 288 pages. ISBN 0-88266-773-4.

Low-Cost Pole Building Construction, by Ralph Wolfe. Includes 10 building plans with elevation drawings and materials lists, 19 examples of already built homes, and detailed descriptions of pole building construction. Paperback. 192 pages. ISBN 0-88266-170-1.

Small-Scale Livestock Farming: A Grass-Based Approach for Health, Sustainability, and Profit, by Carol Ekarius. Presents a natural, organic approach to livestock management that produces healthier animals, reduces feed and health-care costs, and maximizes profitability. Readers will also learn the basics of animal housing, fencing, health, reproduction, and feeding, plus how to market a livestock business using conventional and alternative strategies. Paperback. 192 pages. ISBN 1-58017-162-1.